CSI CHAPTERS

A Guide to Time Travel

by Tony Burnett

Contents

History Is Written in the Stars

Grandpa?

What is it, Teresa?

Is it possible to travel back in time?

No, I don't think so. Most people say it can never be done. Mind you, we can easily *look* back into the past, if we want to. Looking back is easy, but we can't actually go back there ourselves. We just have to look up at the stars to see into the past.

What do you mean?

Well, do you know the name of the nearest star to us, and how far away it is?

Yes, it's the sun. And right now, it's not far away at all, Grandpa. It's just over there behind that cloud!

Good one, Teresa. You have a fine sense of humor, and you're quite right, too! Our sun really *is* the closest star. It's about 93 million miles (150 million kilometers) away from Earth. That's a long way for us if we want to go there, but it is really close compared to any of the other stars we can see. So, after the sun, what is the closest star to Earth, Teresa?

The sun is the only star I know about.

Well, the closest star to Earth, after the sun, is called **Proxima Centauri**. Before I tell you how far away it is, I want to ask you something else. Has anyone ever told you about the speed of light?

The Fastest Thing in the Universe

Don't worry if you are unsure about the speed of light. Let's see if you can figure out the speed for yourself. It's easy. Have you been taught in school how to calculate speed?

Yes. Our teacher said that we just divide the distance we've gone by the time that it took us. Is that right?

Yes. That's exactly right. Scientists know that light takes 8.3 minutes to reach us from the sun. Using your calculator, we can figure out exactly how fast light can go.

Okay. I'll have to divide 93 million miles by 8.3 minutes, right?

Yes. That will give you the speed in miles per minute.

Okay, let's see. I type in 93 million and divide by 8.3. That equals ... Wow!

What is it?

It says 11,204,819! Okay, let's round that number down to make things easier. That means the light is traveling at 11 million miles every minute (18 million kilometers per minute)! Oops. I think I must have made a mistake there, Grandpa. That can't be right, can it?

The calculator is right, Teresa. It's good that you're questioning your answer, too. I like that. Now, there's only one thing left to do. The speed of light is usually given in distance per second, not distance per minute. How can you change your answer to reflect the speed of light in distance per second?

Well, there are 60 seconds in a minute, so all I need to do is divide the answer by 60. Let's see ... I type 11,204,819 and divide it by 60. It comes to 186,746.

To calculate the speed of light in miles (using rounded numbers)

1. 93,000,000 miles ÷ 8.3 minutes = 11,000,000 miles per minute
2. 11,000,000 miles per minute ÷ 60 seconds **= 186,000 miles per second**

To calculate the speed of light in kilometers (using rounded numbers)

1. 150,000,000 kilometers ÷ 8.3 minutes = 18,000,000 kilometers per minute
2. 18,000,000 kilometers per minute ÷ 60 seconds **= 300,000 kilometers per second**

Wow! That's amazing, Grandpa. Light travels at about 186,000 miles per second (300,000 kilometers per second).

That's right. It is fast, isn't it? Light can travel around the whole world *seven* times in the time it takes you to say, "That's amazing." I was the same age as you when I first heard about how fast light is. I can still remember the surprise that I got, too. There is nothing in the whole **universe** that is faster than light. You should think about that the next time you are sitting out in the sun. When you feel the light warming your

skin, you will know that it has traveled 93 million miles (150 million kilometers) in just *eight* minutes! Perhaps it left while you were still inside, busy sneaking Grandma's cake from the refrigerator?

Hey, I didn't think you noticed that. Well, I just can't resist Grandma's cake!

Yes. I like it, too. Anyway, let's go back to Proxima Centauri, shall we?

How Time Flies

Now that you know how fast light really goes, how long do you think it would take a light beam to go from here to Proxima Centauri, the next closest star to Earth after the sun?

I guess it would take the light beam about one day.

No. It's longer than that. Keep guessing.

Longer than a day, okay, let me think. Does it take a week?

No, it's more than that, too. In fact, it's a whole lot more. Believe it or not, light can go from here to the moon in just over one second, from here to the sun in just over eight minutes, and from here to **Neptune** in about four hours. But it takes light more than *four years* to reach Proxima Centauri!

Four years! Wow! Are you sure, Grandpa?

Yes, I'm very sure. Light takes about 4.3 years to reach Proxima Centauri. In terms of light travel, that's not a long time at all. Light takes 30,000 years to go from here to the center of our own **galaxy**, and about 100,000 years to travel across the **Milky Way** from one side to the other.

That's incredible!

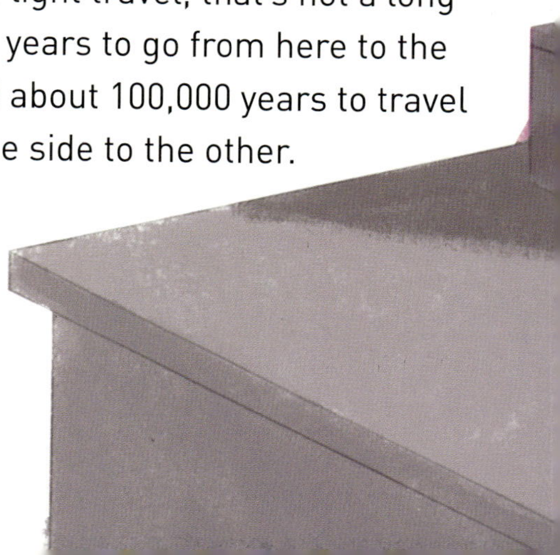

SUN
PROXIMA
CENTAURI
Light takes 4.3 years

Yes, and although we live in the Milky Way, it's only one galaxy among billions of others. The closest other galaxy to us is more than *2 million* light years away, and some of these galaxies are so far away that their light takes more than a billion years to reach us. It really is a big, wide universe out there!

It sure is! It's hard to imagine how big it is, isn't it?

Yes. The stars are a lot farther away than most people think. Try to imagine our **solar system**, from the sun to Neptune, contained within something the size of a marble. If you place the marble in one corner of your classroom, then the closest star would be over in the opposite corner. That should give you some idea of the distance involved.

The stars are far away, and it's the enormous distance between them that lets us look back into the past. We are able to see something only when the light from it reaches our eyes. That can take years, if the object is a long way away. For instance, if you go outside tonight and look up at Proxima Centauri, you will be seeing it as it was more than four years ago. Last night it might have exploded and died, but even if it *did* blow up, we'd still be able to see it shining for the next four years. That's because the light that it gave out before it died comes steadily across space toward us.

The farther away things are, then the farther back we can see. When we look through our big telescopes and see galaxies millions of light years away, it really means that we are looking millions of years back into the past.

Wow! That's so cool. I've never thought of being able to look back into the past like that. It's too bad that we can't go back there ourselves, though. It would be fun to travel back in time.

Past or Future?

Maybe it would be fun to travel into the past, but if I were given a choice, I think I'd rather go into the future.

Why, Grandpa?

Well, I suppose it's just because I'm curious. We already know about things that have happened in the past, but nobody knows what the future will be like. If I had the chance, I think I'd like to go on ahead and see what is going to happen.

I think I'd like to do that, too.

Well, perhaps we can go off together some day! If we do, then I'll try to bring one of those cakes with me, unless I can persuade Grandma to join us, of course. Going into the future might be hungry work! Hungry or not, though, it will always be a one-way trip. If we do go off at high speed, then there's no way we'll ever be able to come back to our own time again. Once we're *in* the future, then I'm afraid we're going to have to stay there. We'll have to be quite sure that we want to go before we leave, won't we?

Yes, but it would be really exciting, wouldn't it? I wish we could go.

Well, there actually *is* a way we can go into the future. It's not going to be easy, though.

There is a way? I can't believe it! How is that possible, Grandpa?

Back in 1905, a man named **Albert Einstein** proved that the flow of time depends on how fast we ourselves are moving.

What do you mean?

Well, believe it or not, the faster we travel, the slower time passes for us compared to other people who aren't traveling as fast. This is not just something that *seems* to happen, it really *does* happen! For instance, if we could go fast enough on a spaceship, then we might be away for two of our years, then find out that ten or twenty years had passed on Earth while we were gone. If we went *really* fast, then we might come home to find that our children were older than we were!

Oh, come on, Grandpa. That can't be true! How could a child possibly be older than his or her parents?

Well, I know it sounds silly, but it really is true. If we go fast enough, and we spend a year on the spaceship, our children could have aged by 30 or 40 years while we were away. How old are you now, Teresa?

I turned ten last month.

Okay. Well I know that your mother is 36 this year, so let's pretend that she goes away on the spaceship and that you stay behind, shall we?

Okay.

Well, if she went to space for a year, and was 37 when she came back again, you could be 40, or even older, if the spaceship traveled fast enough. Your mother would be surprised, wouldn't she? She'd also have a lot of catching up to do. She'd come home to find that you'd finished school, grown up, and maybe started a family of your own. All of this would happen while she was away for only *one* of her years. For her, stepping out of the spaceship would be stepping 30 years into the future. And you never know, do you? She might come home to find she has a clever young granddaughter, just like you, waiting to meet her grandma for the first time!

I don't know about that, Grandpa! Are you sure that would happen? Me growing older faster than my own mother, I mean. It just doesn't seem possible.

I know. It seems ridiculous even to talk about it. Trust me, the idea sounds so crazy that many scientists have tested and retested Albert Einstein's work. Yet his theory has never failed, not even once. There is no doubt at all that he was right. What we are talking about is something called **time dilation**. Mind you, we'd have to be going *very* fast before we'd be able to notice anything.

How to Calculate Time Travel

How quickly would we have to travel to notice the effects of time dilation, Grandpa?

Well, the faster the better, really. It would be great if we could get close to the speed of light itself, which is about 93 million miles per hour (150 million kilometers per hour), but that's *far* faster than any of our spacecrafts can go. Still, you never know. One day, we may be able to do it. Even at half the speed of light, or about 46 million miles per hour (75 million kilometers per hour), we would soon notice the difference in time.

What do you mean? How much difference?

Well, you can use the calculator again and figure it out. It's not difficult.

Okay. What do I have to do?

Well, there are five simple steps. The first is to choose a speed that is some **fraction** of the speed of light, such as one-half, one-quarter, or three-quarters. Then, write the fraction as a **decimal**, square it, subtract the answer from one, and take the square root. That's all there is to it. Has your calculator got a square root button? Look for the square root symbol on the keypad.

Yes, here it is.

Good. Well, let's try one. How fast should we go?

How about we go at half the speed of light? Is that okay?

Sure, that's fine, Teresa. A half is easy, because it comes to 0.5 when we write it as a decimal. Now, let's try the five steps I mentioned.

Okay.

1. Teresa has decided to travel at half ($\frac{1}{2}$) of the speed of light.
2. She writes the fraction as a decimal number ($\frac{1}{2} = 0.5$).
3. She squares the decimal number ($0.5^2 = 0.25$).
4. She subtracts the answer from 1 ($1 - 0.25 = 0.75$).
5. She takes the square root of the answer ($\sqrt{0.75} = 0.8660$).

The calculator shows 0.8660, Grandpa. I think that's right, but what does the 0.8660 stand for?

It means that, for every year on Earth, only 0.8660 of a year will go by on the spaceship. Let's see. It might be easier if we multiply by ten, I think. If we do, then the answer tells us that for every ten years that go past on Earth, only 8.660 years will pass for us on the ship. Okay, that's a difference of 1.34 years, which is about one year and four months, or sixteen months.

All right, but I'm still not too sure about it, Grandpa.

Let's suppose that you had a twin ten-year-old brother, and that you went off on the ship while he stayed at home. If the ship traveled at half the speed of light and came back on your twentieth birthday, you would be one year and four months, or sixteen months, younger than him. In other words, your twin brother would be twenty, and you would be only eighteen years and eight months old!

Well, I guess it must be right, but it just doesn't seem real somehow! What happens if we go even faster than half the speed of light?

Teresa's Calculation

The faster we go, the bigger the difference. I know it takes a little bit of getting used to, but don't be scared of trying a few calculations for yourself. It seems really strange at first, but once you get the hang of it, you can have a lot of fun by trying a few different speeds and figuring out what the time difference would be. How would you like to try a faster one?

Okay. Can we go at nine-tenths the speed of light this time?

Of course we can. And that's also a good choice. Nine-tenths is another easy one because it just comes to 0.9 when we write it as a decimal. Okay, why don't you use 0.9 and try it yourself?

1. Teresa has decided to travel at nine-tenths ($\frac{9}{10}$) the speed of light.
2. She writes the fraction as a decimal number ($\frac{9}{10} = 0.9$).
3. She squares the decimal number ($0.9^2 = 0.81$).
4. She subtracts the answer from 1 ($1 - 0.81 = 0.19$).
5. She takes the square root of the answer ($\sqrt{0.19} = 0.4359$).

Okay, my answer is 0.4359, Grandpa. Is that right?

Correct, Teresa! That was really easy, wasn't it? Now, can you tell me what it means?

Well, I think it means that, for every ten years on Earth, only a little over four years goes by on the spaceship. Is that right?

That's exactly what it means, and you'd surely notice the difference between yourself and your brother this time. When you came back home, your twin brother would be twenty and you would be only fourteen.

Hmm. I don't think I'd like that much, Grandpa. Brothers are annoying enough as it is without having a twin who's six years older!

I agree! But building a spaceship that can approach the speed of light is not going to be easy. We are nowhere near it yet, and some people think we may never be able to do it.

Oh, I hope we can. How fast do our spaceships go now?

Well, I think the fastest ones are those that we've sent away from our solar system and out into deep space. They have a speed of about 30 miles per second (48 kilometers per second), and that's about 6,000 times slower than light. At the moment, comparing our spaceships' speed to light speed is a little bit like comparing the speed of a snail to a race car!

Does this mean our spaceships are too slow for us to see any time difference?

Yes, they're much too slow. There *is* a time difference, of course, but it's too small for us to see. The difference in age between a baby born on a spaceship going at 30 miles per second (48 kilometers per second) and one born at the same time here on Earth works out to be only about 20 seconds in 50 years. It's a different story at high speeds, though.

Grandpa's Amazing Calculation

Let's try another speed. You can go as fast as you like, as long as it's slower than the speed of light.

Okay. So how about this time we go only a tiny bit slower than the speed of light? Something like one ten-thousandth less than light speed.

Okay. That is 0.9999 times the speed of light. Calculate it, then see what happens.

The calculator shows 0.01414, Grandpa. Does that mean only 0.01414 of a year goes past on the spaceship while a whole year goes past on Earth?

That's right. If you multiply your answer by 365, you'll find that 0.01414 of a year comes to a little more than five days. In other words, for every five days that you spend on the spaceship, your twin brother will age by one full year back here on Earth. If you like, you can think of it another way.

For every year you spend on the spaceship, your brother would age by 73 years back on Earth! If you were away on a journey that took ten of your years on the spaceship, then you'd find that Earth would be more than *700 years older* when you came back home.

Wow! Seven hundred years older! But that's not good. All my friends would have grown old and died before I got back!

Back to the Future

So, here's the amazing thing, Teresa. If you took off on your next birthday and were away on a spaceship for ten years, traveling at almost the speed of light, just as we calculated, you would arrive back home and be 700 years into the future! None of your family or friends could possibly be there to welcome you back. There's another thing, too. Earth would've changed. It would be very different compared to what it was when you left. That's why I said before that going into the future is a one-way trip.

You know, Grandpa, I wouldn't like to lose all my friends like that. Now I'm not too sure that I'd really want to go. I'd never see you, Grandma, or my other family and friends again, either!

Don't worry – you might be able to take your family and friends along with you. That way you could all share the adventure. Think of the things you'd see! You could fly to the stars!

Wow! That does sound cool. I'd love to go if everyone could come with me, especially you and Grandma!

Well, I think it might be a bit too late for Grandma and me. I don't think she would ever want to leave home, and I'd never go without her. Flying to the stars is going to be something for young people like you to do, not for old folks like us. If it ever does happen, then it will be the greatest

voyage of exploration that humans have ever achieved. Let's hope that one day you can be part of it. There will be lots of new sights to see and many new planets to explore. And, who knows? If you're extra lucky, you might even find yourself a world where pieces of cake grow on trees!

That would be great, Grandpa! If there really are cake trees, I hope they grow cakes that taste as good as Grandma's!

Glossary

Albert Einstein – a brilliant scientist who was born in Germany in 1879 and died in the United States in 1955. Voted "Man of the Century" in 1999, he became famous for his discovery of time dilation and for the theory of relativity.

decimal[2] – a number written using the base 10

fraction[2] – a quantity less than a whole, or less than one

galaxy – a large family of stars held together by gravity. In the early 1920s, Edwin Hubble discovered that the Milky Way was not the only galaxy in the universe, and we now know that there are more than 100 billion others.

Milky Way – the name given to our own galaxy. The Milky Way contains about 200 billion stars and is so big that light takes 100,000 years to go from one side to the other.

Neptune – the eighth planet in our solar system

Proxima Centauri *(prok-suh-muh sen-tor-ree)* – the closest star to our own sun

solar system[3] – the name given to our own family of planets in orbit around the sun. Starting from the closest to the sun, the planets include Mercury, Venus, Earth, Mars, Jupiter, Saturn, Uranus, and Neptune.

time dilation – the slowing of time experienced by objects going very fast compared to those that are not

universe – the whole of space, no matter how far we go in any direction. The universe contains all of the energy, galaxies, stars, and planets that are possible.

Academic Vocabulary Key	**4** Economics	**8** US History	**12** Technology
1 English Language Arts	**5** Civics	**9** World History	**13** General Arts
2 Mathematics	**6** Geography	**10** Health	**14** Dance/Music
3 Science	**7** General History	**11** Physical Education	**15** Theater/Visual Arts